S. Naresh
C.N.Hari Prasath

Wild Edible Mushroom in Forest Ecosystem

S. Naresh
C.N.Hari Prasath

Wild Edible Mushroom in Forest Ecosystem

LAP LAMBERT Academic Publishing

Impressum / Imprint
Bibliografische Information der Deutschen Nationalbibliothek: Die Deutsche Nationalbibliothek verzeichnet diese Publikation in der Deutschen Nationalbibliografie; detaillierte bibliografische Daten sind im Internet über http://dnb.d-nb.de abrufbar.

Bibliographic information published by the Deutsche Nationalbibliothek: The Deutsche Nationalbibliothek lists this publication in the Deutsche Nationalbibliografie; detailed bibliographic data are available in the Internet at http://dnb.d-nb.de.

Coverbild / Cover image: www.ingimage.com

Verlag / Publisher:
LAP LAMBERT Academic Publishing
ist ein Imprint der / is a trademark of
OmniScriptum GmbH & Co. KG
Heinrich-Böcking-Str. 6-8, 66121 Saarbrücken, Deutschland / Germany
Email: info@lap-publishing.com

Herstellung: siehe letzte Seite /
Printed at: see last page
ISBN: 978-3-659-51803-4

Wild Edible Mushroom in Forest Ecosystem

By

S.Naresh

C.N.Hari Prasath

2014

ACKNOWLEDGEMENT

I express my sincere and eternal gratitude and respect to my chairman of seminar work, **Dr. Dr.P.Muthulakshmi, Ph.D.,** Assistant Professor (Department of Forest Products and Utilisation) Forest College and Research Institute, Mettupalayam, for his valuable guidance, constant encouragement throughout the course of this investigation and untiring helps without which this work would not have attained its shape.

My special thanks to **Mr.G.Chakravarthi** and **Mr.C.Veeramani** for helping me in preparing the seminar work for bringing out in successful manner.

"A friend is one who will feel the love of his friend wherever he is....." in this sense I thank all my **Spartanz** for their kind love, care and timely help during my seminar work. I equally thank my lovable **seniorz and juniorz** for giving energy to me during all time.

On a personal note, I wish to express my gratitude to my father **Thiru Subbian,** who is the reason for my upliftment from the childhood till now. And I wish to express my respect to my mother **Tmt. S.Indra** and my beloved sister **Mrs.S.Naveena** for their sacrifice, love, care, motivational and moral support during the course of study.

Last, but not the least I thank **"THE ALMIGHTY"** for his blessings.

(S.Naresh)

ABSTRACT

Mushrooms are seasonal fungi, which occupy diverse niches in nature in the forest ecosystem. They predominantly occur during the rainy season and also during spring when the snow melts. Mushrooms are in fact the 'fruit' of the underground fungal mycelium. Mushrooms are sources of food for wildlife and fungi that cause decay in living trees are beneficial to many species of birds and mammals. It act as a source of food, income and to maintain the health of forest. Wild fungi also have medicinal properties, some of which are found in edible species. Wild useful fungi therefore contribute towards diet, income and human health. Many species also play a vital ecological role through the symbiotic relationships known as mycorrhizal that they form with trees.

Characterization of wild edible mushroom described for the morphological characters like are colour, size, shape, odour, texture, accurate and consistent notation of sporocarps colour. Important wild edible mushroom in existing forest ecosystem are Tropical Mushroom, Subtropical Mushroom, Moist and Dry Deciduous Forest, Shola forest, Thorn wild edible mushroom, Coniferous Forest Ecosystem. Drying and preservation fresh wild edible mushrooms have a short period during which they can be eaten or consumed. Owing to their perishable nature, they quickly deteriorate, rot, or shrivel up. On questioning local people about this aspect, it was realized that they consumed large number of the mushrooms in fresh form and only a few are preserved after sun-drying, smoke drying or salting.

Since thousands of years, edible fungi have been revered for their immense health benefits and extensively used in folk medicine. Specific biochemical compounds in mushrooms are responsible for improving human health in many ways. The uses of the wild edible mushrooms are good for heart, low calorie food, prevents cancer, anti-aging property, regulates digestive system and strengthens immunity. Sharpening our knowledge will be a good investment and is an essential part of maintaining continuous production of wild edible mushrooms for future generation.

CONTENT

Sl.No	Title	Page No
1	Introduction	09
2	Pre-history of wild edible mushroom used by primitive people	11
3	Difference between edible and non-edible mushroom	12
4	Nomenclature for macro fungi	13
5	Characteristics and life cycle of wild edible mushroom	13
6	Important wild edible mushroom in existing forest ecosystem	15
7	Domestication and cultivation of wild edible mushroom in India	30
8	Status and scenario of mushroom cultivation in India	34
9	Drying and Preservation methods of wild edible mushrooms	38
10	Marketing and production of wild edible mushroom in India	41
11	Uses of wild edible mushroom	44
12	Constraints in wild edible mushroom	47
13	Conclusion	48
	Reference	49

Wild Edible Mushroom in Forest Ecosystem

1. Introduction

Mushrooms are seasonal fungi, which occupy diverse niches in nature in the forest ecosystem. They predominantly occur during the rainy season and also during spring when the snow melts. Mushrooms are in fact the 'fruit' of the underground fungal mycelium. They are macromycetes forming macroscopic fruiting bodies such as agarics, boletes, jelly fungi, coral fungi, stinkhorns, bracket fungi, puffballs and bird's nest fungi. They are fleshy, sub fleshy, or sometimes leathery and woody and bear their fertile surface either on lamellae or lining the tubes, opening out by means of pores. The lamellate members are called agarics and the tube bearing poroid members, as boletes and polypores. Among fungi, Basidiomycotina in particular have attracted considerable attention as a source of new and novel metabolites with antibiotic, antiviral, phytotoxic and cytistatic activity. Mushrooms alone are represented by about 41,000 species, of which approximately 850 species are recorded from India mostly belonging to Agaricales, also known as gilled mushrooms (for their distinctive gills), or euagarics. The order has 33 extant families, 413 genera and over 13000 described species.

Fungi are considered the largest biotic community after insects (Sarbhoy *et al.* 1996). Yet only a small fraction of total fungal wealth has been subjected to scientific scrutiny and the new generation hi tech mycologists have to unravel the unexplored and hidden wealth. Out of 1.5 million fungi around the globe, only 50% are characterized until now and one third of total fungal diversity of the globe exists in India (Butler and Bisby 1960; Bilgrami et al. 1981; 1991; Manoharachary 2002; Manoharachary et al. 2005). Mushrooms comprise largely the group of fleshy fungi, which include bracket fungi, fairy clubs, toadstools, puffballs, stinkhorns, earthstars, bird's nest fungi and jelly fungi. Generally they live as saprophytes however some are serious agents of wood decay.

All types of mushrooms are important in decomposition processes, because of their ability to degrade cellulose and other plant polymers. Besides they serve as natures trash burners and soil replenishers and thus help in rejuvenating the ecosystem. Ample species of wild edible and medicinal mushrooms occur in all the biodiversity rich regions during the rainy season. India being the top ten mega diversity has innumerable mushroom species and their ethno mycological importance. The wood of living or dead trees, or the leaf litter or the soil produce mushroom through the branching mycelial infiltration some mushrooms are found growing in association with trees of a particular family or genus.

Though India has rich macro fungal biodiversity, most traditional knowledge about mushrooms come from the far East countries like China, Japan, Korea, Russia where mushrooms like *Ganoderma, Lentinus, Grifola* and others were collected and used since time immemorial. Most of the mushrooms grow abundantly in nature and their commercial harvest is being undertaken for benefit in these countries. Recent reports show a tradition of wild mushroom picking, their consumption and sale in the market in countries like Mexico, Italy, Australia and many others (Arora 2008, Guzman 2008, Sitta and Floriani 2008). However, the ecological data available on some of the taxa is still not enough and systematics of wild mushrooms has received more attention than other threatened aspects like conservation.

1.1. How it`s feed?

Dependent on dead and living material for their growth, they obtain their nutrient in three basic ways

1. Saprobic – growing on dead organic matter
2. Symbiotic – growing in association with other organism
3. Pathogenic (or) Parasitic – causing harm to other organisms

Majority of wild edible fungi species are symbiotic and form mycorrhizae with trees.

1.2. Wild Edible Fungi Importance

Fungi are important organisms that serve many vital functions in forest ecosystems including decomposition ,nutrient cycling, symbiotic relationships with trees and other plants, biological control of other fungi, and as the causal agents of diseases in plants and animals. Mushrooms are sources of food for wildlife and fungi that cause decay in living trees are beneficial to many species of birds and mammals

1. As a source of food
2. As a source of income
3. To maintain the health of forest

2. Pre-history of wild edible mushrooms used by primitive people

Wild edible fungi (WEF) have been collected and consumed by people for thousands of years. The archaeological record reveals edible species associated with people living 13 000 years ago in Chile (Rojas and Mansur, 1995) but it is in China where the eating of wild fungi is first reliably noted, several hundred years before the birth of Christ (Aaronson, 2000). Edible fungi were collected from forests in ancient Greek and Roman times and highly valued, though more by high-ranking people than by peasants (Buller, 1914). Caesar's mushroom (*Amanita caesarea*) is a reminder of an ancient tradition that still exists in many parts of Italy, embracing a diversity of edible species dominated today by truffles (Tuber spp.) and porcini (*Boletus edulis).*

China features prominently in the early and later historical record of wild edible fungi. The Chinese have for centuries valued many species, not only for nutrition and taste but also for their healing properties. It is less well known that countries such as Mexico and Turkey, and major areas of central and southern Africa also have a long and notable tradition of wild edible fungi. The list of countries where wild fungi are reported to be consumed and provide income to rural people is impressive. The threat posed by poisonous and lethal species is often overstated.

Incidents of poisoning and deaths are few and far between compared to the regular and safe consumption of edible species, but publicity and cultural attitudes continue to fuel an intrinsic fear of wild fungi in some societies. This is more commonly found in developed countries and has undoubtedly led to general beliefs that global use of wild edible fungi is small-scale and restricted to key areas. As this publication conclusively shows, this is simply not true the use of wild edible fungi is both extensive and intensive, though patterns of use do vary

Wild edible fungi add flavour to bland staple foods but they are also valuable foods in their own right. Local names for termite mushrooms (Termitomyces) reflect local belief that they are a fair substitute for meat, a belief that is confirmed by nutritional analyses. Not all wild edible fungi have such high protein content but they are of comparable nutritional values to many vegetables. Wild edible fungi are sold in many local markets and commercial harvesting has provided new sources of income for many rural people.

Wild fungi also have medicinal properties, some of which are found in edible species. Wild useful fungi therefore contribute towards diet, income and human health. Many species also play a vital ecological role through the symbiotic relationships known as mycorrhizal that they form with trees.

3. Differences between Edible and Non-Edible mushroom

Sl.No	Edible Mushroom	Poisonous Mushroom
1	Possess a flat ,rounded or depressed cap	Possess a pointed cap
2	Presence of worms or insects	Absence of worms or insects
3	Dull coloured	Brightly coloured
4	Absence of warts or scales on the surface of cap	Presence of warts or scales on the surface of cap
5	Pleasant odour	Unpleasant odour
6	Colour of fruiting bodies does not turn dark or breakage	Colour of fruiting bodies turns dark on breakage (black or purple)
7	Using silver spoon there is no change in the color	Using silver spoon there is change in the color to black
8	Cooking with onion there is no color change in onion	Cooking with onion there is color change in onion to black.
9	**Examples:** *Agaricus bisporus, P. osteratus P. florida, Calocybe indica.*	**Examples:** *Ganoderma applantum, Polyporus pargamenus,Amanita muscaria*

S.R. Joshi (2011)

4. Nomenclature for Wild Edible Mushroom

Macrofungi normally grow under humid conditions and flourish only when ecological/edaphic conditions are favorable. It is usual for a particular fungus to produce visible fruiting body/sporocarps under a precise combination of conditions (e.g. geographic location, elevation, temperature, humidity, light and surrounding flora). Their mycelia exist in soil, humus, plant litter, decaying wood or other substrata inconspicuously for a very long period. Macrofungi are the agents responsible for the disintegration of organic matter, important component of forest ecosystems and play a vital role in ecosystem dynamics (litter decomposition, nutrient cycling and nutrient transport).

They prefer grassy grounds (lawns, gardens, pastures, hilltops, meadows and fields), forests (evergreen, moist deciduous and semi-evergreen), jungles, shoals, sacred grooves, orchards, marshes, swaps, sand dunes, dung (domestic and wild), burnt ground, richly manured ground, mossy rocks and soil (rocky, alluvial and acidic soils). However, they also grow along streets, road sides and rail roads, cellars, mines, roofing tiles are often ideal for growth of some species. There are different names given to macrofungi that signifies their habitat by Purkaystha and Chadra (1985) *viz.*

1. Species grown in grasslands are known as particolous Eg. *Lycoperdon maximum*
2. Species grown in wood land are known as silvicolous Eg. *Russula* spp
3. Species grown in wood, tress, twigs and stump are known as lignicolous Eg. *Lentinus* spp and *Pleurotus* spp

5. Characteristics and life cycle of wild edible mushroom

Characterization of wild edible mushroom described for the morphological characters like are

a) Colour

b) Size

c) Shape

d) Odour and texture

e) Accurate and consistent notation of sporocarps colour

5.1. Parts of Mushroom

1) Pileus

The pileus (cap) is the expanded part of the mushroom. It is quite thick and fleshy in consistency, more (or) less round (or) convex on the upper side and usually white in colour. The surface is generally smooth.

2) Gills

On the underside of the pileus are radiating plates the gills which bear the spores.

3) Stipe

The Stipe (stem) is attached to the pileus in the centre. It is cylindrical, white fleshy, smooth, quite firm and compact and varies in length.

4) Ring:

Annulus (or) ring is usually present in the matured *Agaricus campestris* around the upper end of the stem. It is joined to the stem, very delicate easily ribbed off or may be even washed during rains.

5) Volve:

The deadly *Amanita phalloides* has a cup like structure attached to the lower end of the stem, from which the stem appears to spring. This volva is also called as 'death cap' (or) 'poison cap'.

5.2. Life Cycle:

Mushroom spores germinate under optimum temperatures and moistures to form a mass of mycelium. Spores germinate to form the primary mycelium which is haploid and uninucleate. Mycelia from different spores fuse to form the secondary mycelium which is septate.

Fragments of this mycelium are complete and are capable of independent growth. The mycelium produces enzymes that digest carbohydrates, lipids, and protein which can be absorbed by the hyphae. This is the spawn stage where energy is stored until the fruiting bodies emerge in reproduction. This stage is the formation of the visible mushrooms which starts as pinheads or primordia and develops into the button stage. The mushroom develops a cap and forms gills underneath. Basidia cells are formed at the edges of the gills which later produce basidiospores which are released to start the cycle again.

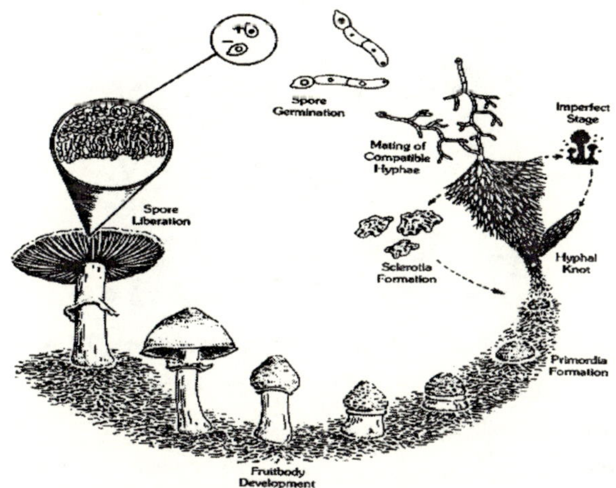

6. Important wild edible mushroom in existing forest ecosystem

6.1. Tropical Mushroom

Lycoperdon maximum - **Wolf-fart puffball**

Fruiting body-1-4.5 cm across, solitary, globose or pyriform, mealy granules cover the exoperidium, endoperidium greyish brown, smooth, whitish to greyish brown;

angiocarpic, gleba white, soft, fleshy at young stage, brown cottony at later stage; spores olive brown, spores 2.5-4.5μm. Edible when young.

Pleurotus ostreatus – **The pearl oyster mushroom**

Sporophores usually growing in clusters on dead tree trunk or branches and rarely on living trees, usually hygrophanous, whitish, large, tough, when old. Pilus 8.0-15. Cm or more broad. Spathulate to kidney shaped, white, grey or sometimes yellowish after drying Surface smooth, margin incurved. Gills crowded, decurrent, white, yellow when dry, broad. Stipe eccentric or lateral, 1.0-3.0 cm long, 0.5-2.0 cm thick, firm sometimes hairy at the base. Hymenophoral trama irregular, Basidia 4-spored, 30.0-38.0x6.0 μm. Basidiospores white, oblong, 7.0-10.0 μm long. Spore print lilac.

Coprinus comatus - **The shaggy mane mushroom**

Members of the genus *Coprinus* have been collectively known as the "inky caps" because of the curious character of autodeliquescence of their gills-- in other words self-digestion to release their basidiospores. Most of these species have gills that are very thin and very close to one another, which does not allow for easy release of the spores. Moreover the elongated shape of mushroom does not allow for the spores to be shot off the basidia and get caught in air currents as in most other mushrooms. *Coprinus* species "compensate" for this by a sequential maturation of the spores from the bottom of the gill towards the top. After the spores have matured and been released, the gill tissue digests itself and begins to curl up, allowing easy release of the basidiospores above.

In other words, the digestion opens up the fruiting body so that spores from further up the gills become exposed to the air and a clear path of spore release. The self-digestion continues until the entire fruiting body has turned to black ink.

Volvariella speciosa – **Straw mushroom**

Sporophores usually growing on rotten paddy straw usually distinguished by its absence of pigmentation. Pileus 8.0-10.0 cm in diameter, hemispherical, sub-fleshy. Gills distant, white, ventricose. Stipe 6.0-8.0 cm longs, narrowed upward, 2.5-3.0 cm broad at the base, 0.8-1.0 cm at the top, solid, volva bilobed, brown, descending, margin curved, smooth. Hymenophoral trama inverse. Basidiospores pink. Spore print pink.

Agaricus bitorquis – **Button Mushroom**

Found scattered on pastures, lawns and on scattered manure. Pileus: 5-10 cm diameter, convex in young, flattened in old fruit bodies. White to pale-brown, finely scaly surface, margin entire. Gills: prominent, crowded free, white in young, pink to dark brown to blackish in older ones. Stipe: central, equal, 4-10 cm long, 1-2 cm thick, white to pale brown, annulate, annulus white membranous prominent. Basidia:2 spored, spores brown, ellipsoid, 7x5.5 µm2. Spore print: sepia to brown.

6.2. Subtropical Mushroom

Pleurotus platypus – **Meadow mushroom**

Sporophores solitary or groups occurring on decaying dead tree trunk or branches usually hygrophanous, whitish, large, tough, when old. Pilus 10.0-15.0 cm or more broad, spathulate to kidney shaped, white, grey. Surface smooth, margin incurved. Gills decurrent, white, yellow when dry, broad. Stipe eccentric or lateral, 1.0-

3.0 cm long, 0.5-2.0 cm thick. Hymenophoral trama irregular, Basidia 4-spored, 30.038.0x6.0 µm. Basidiospores hyline, ellipsoid or cylindrical, 6.0-12.0 µm long. Spore print whitish.

Pleurotus djamor – **Oyster mushroom**

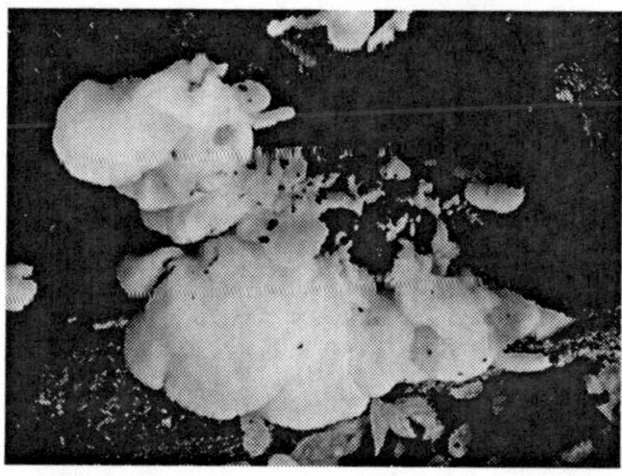

Pleurotus are white-rot fungi on hardwood trees, although some also decay conifer wood Gills: Running down the stem; close or nearly distant; whitish. Stem: Sometimes absent or rudimentary, but often present; 1-7 cm long and up to 1.5 cm thick; eccentric or lateral--or central. Flesh: Thick; white. Odour and Taste. Odour distinctive but hard to describe ("like oyster mushrooms" works well, but makes for a circular description); taste mild. Chemical Reactions: KOH on cap surface orangish.

Auricularia auricula - **Jew's ear mushroom**

The common name for this fungus, the "Jelly Ear," is very appropriate--a rarity with common names in English. Just take a look at the photos to the right, and you can see why people think Auricularia auricula looks like the missing part of a Van Gogh self portrait. Unlike the cup fungi, this fungus has jelly-like flesh, and its spores are catapulted from little spore-holders, placing it in the Basidiomycetes rather than in the Ascomycetes, where spores are forcibly shot out of little spore-jets.

Ecology: Saprobic on decaying hardwood and conifer sticks and logs; spring, summer, and fall (sometimes even in winter); widely distributed in North America.

Fruiting Body: Wavy and irregular; typically ear-shaped; 2-15 cm; gathered together and attached at a central or lateral position; fertile surface (usually the "downward" one) gelatinous, tan to brown; sterile surface (usually the "upper" one) silky to downy, veined, irregular, brown; flesh thin, gelatinous-rubbery.

Spore Print: White.

Agrocybe aegerita –**The chestnut mushroom**

Agrocybe Aegerita is a species of mushroom that belongs to the white rot fungi but looks like the button mushroom, only darker. It has gills whose colour ranges from pink to dark brown. Hence the mushroom is sometimes referred to as 'brown cap mushroom'. It behaves almost the same as some other species of mushroom like the *Tricholomamatsutake*

Agaricus campustris – **The field mushroom**

The cap is white, may have fine scales, and is 5 to 10 centimeters (2.0 to 3.9 in) in diameter; it is first hemispherical in shape before flattening out with maturity. The gills are initially pink, then red-brown and finally a dark brown, as is the spore print. The 3 to 10 centimeters (1.2 to 3.9 in) tall stipe is predominantly white and bears a single thin ring. The taste is mild. The white flesh bruises slightly reddish, as opposed to yellow in the inedible (and somewhat toxic) Agaricus xanthodermus and similar species. The spores are 7–8 micrometers (0.00028–0.00031 in) by 4–5 micrometers (0.00020 in) ovate.

6.3. Moist and Dry Deciduous Forest

Pearl Oyster mushroom - ***Pleurotus osteratus***

Sporophores usually growing in clusters on dead tree trunk or branches and rarely on living trees, usually hygrophanous, whitish, large, tough, when old. Pilus 8.0-15. Cm or more broad. Spathulate to kidney shaped, white, grey or sometimes yellowish after drying Surface smooth, margin incurved. Gills crowded, decurrent, white, yellow when dry, broad. Stipe eccentric or lateral, 1.0-3.0 cm long, 0.5-2.0 cm thick, firm sometimes hairy at the base. Hymenophoral trama irregular, Basidia 4-spored,30.0-38.0x6.0 µm. Basidiospores white, oblong, 7.0-10.0 µm long. Spore print lilac.

Volvariella volvacea – **Straw mushroom**

Sporophores usually growing solitary or gregarious on rotten organic maters or rotten paddy straw heaps; centrally stipitate. Pileus usually 5.0-12.0 cm in diameter, often larger, campanulate at first, later becoming umbonate, usually grayish sepia but sepia near the umbo as well at the margin, presence of distinct radial sepia colored streak up to the middle of the pileus, soft and fatty to touch, margin sometimes split.

Gills crowded, distinctly formed, free, thin, flesh colored with reddish tinge at maturity. Stipe central, cylindrical, attenuated upward, 8.0-14.0 cm long, whitish, ending below with a solid bulbous base, volva well developed and membranous with margin free. Basidia clavate, tetrasterigmatic, 18.7-28.9x6.8-11.9 µm. Basidiospores oval to ovoid, smooth, thin walled, 6.8-9.3x4.6-6.3 µm. Spore print salmon pink colored on white paper. Pleurocystidia lanceolate to clavate, 47.6-62.9x13.6-18.7 µm.

Volvariella diplasia – **Straw mushroom**

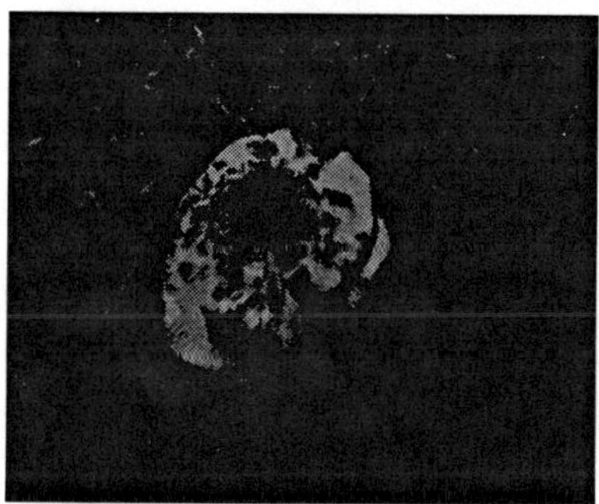

Sporophores usually growing solitary on rotten organic maters or rotten paddy straw heaps; centrally stipitate. Pileus usually 6.0-10.0 cm in diameter, often larger, campanulate at first, later becoming umbonate, usually Light grayish sepia but sepia near the umbo as well at the margin, presence of distinct radial sepia colored streak of the pileus, soft and fatty to touch, margin sometimes split. Gills crowded, distinctly formed, free, thin, flesh colored with reddish tinge at maturity. Stipe central, cylindrical, attenuated upward, 8.0-14.0 cm long, whitish, ending below with a solid bulbous base, volva well developed and membranous with margin free.

6.4. Shola forest

Agaricus bisporus - **Button mushroom**

Found scattered on pastures, lawns and on scattered manure. Pileus: 5-10 cm diameter, convex in young, flattened in old fruit bodies. White to pale-brown, finely scaly surface, margin entire. Gills: prominent, crowded free, white in young, pink to dark brown to blackish in older ones. Stipe: central, equal, 4-10 cm long, 1-2 cm.

Pleurotus spp - **Oyster mushroom**

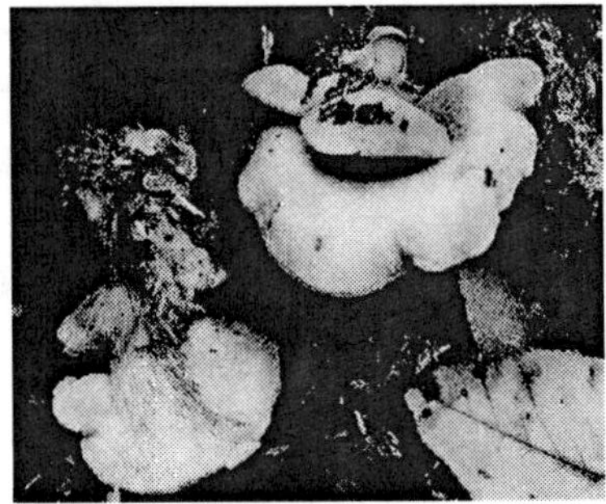

Sporophores solitary or groups occurring on decaying dead tree trunk or branches usually hygrophanous, whitish, large, tough, when old. Pilus 10.0-15.0 cm or more broad, spathulate to kidney shaped, white, grey. Surface smooth, margin incurved. Gills decurrent, white, yellow when dry, broad. Stipe eccentric or lateral, 1.0-3.0 cm long, 0.5-2.0 cm thick. Hymenophoral trama irregular, Basidia 4-spored, 30.038.0x6.0 µm. Basidiospores hyline, ellipsoid or cylindrical, 6.0-12.0 µm long. Spore print whitish.

Polyporus spp – **Bracket mushroom**

The fungal individual that develops the fruiting bodies we identify as polypores resides in soil or wood as mycelium. Polypores are often restricted to either deciduous (angiosperm) or conifer (gymnosperm) host trees. Some species depend on a single tree genus (e.g. *Piptoporus betulinus* on birch, *Perenniporia corticola* on Dipterocarps). Forms of polypore fruiting bodies range from mushroom-shaped to thin effused patches that develop on dead wood. Perennial fruiting bodies of some species growing on living trees can grow over 50 years old. In fruiting bodies with a cap (pileate fruiting bodies) the tissue between upper surface and the pore layer is called context.

6.5. Thorn wild edible mushroom

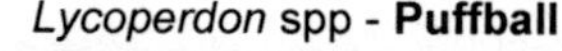
Lycoperdon spp - **Puffball**

Fruiting body-1-4.5 cm across, solitary, globose or pyriform, mealy granules cover the exoperidium, endoperidium greyish brown, smooth, whitish to greyish brown;angiocarpic, gleba white, soft, fleshy at young stage, brown cottony at later stage; spores olive brown, spores 2.5-4.5µm. Edible when young

Pleurotus spp - **Oyster mushroom**

Sporophores solitary or groups occurring on decaying dead tree trunk or branches usually hygrophanous, whitish, large, tough, when old. Pilus 10.0-15.0 cm or more broad, spathulate to kidney shaped, white, grey. Surface smooth, margin incurved. Gills decurrent, white, yellow when dry, broad. Stipe eccentric or lateral, 1.0-3.0 cm long, 0.5-2.0 cm thick. Hymenophoral trama irregular, Basidia 4-spored, 30.038.0x6.0 µm. Spore print whitish.

Volvariella spp - **Straw mushroom**

Sporophores usually growing on rotten paddy straw, usually distinguished by its absence of pigmentation. Pileus 8.0-10.0 cm in diameter, hemispherical, sub-fleshy. Gills distant, white, ventricose. Stipe 6.0-8.0 cm longs, narrowed upward, 2.5-3.0 cm. Broad at the base, 0.8-1.0 cm at the top, solid, volva bilobed, brown, descending, margin curved, smooth. Hymenophoral trama inverse. Basidiospores pink. Spore print pink.

6.6. Coniferous Forest Ecosystem

Laccinum scabrum – **Birch boletes**

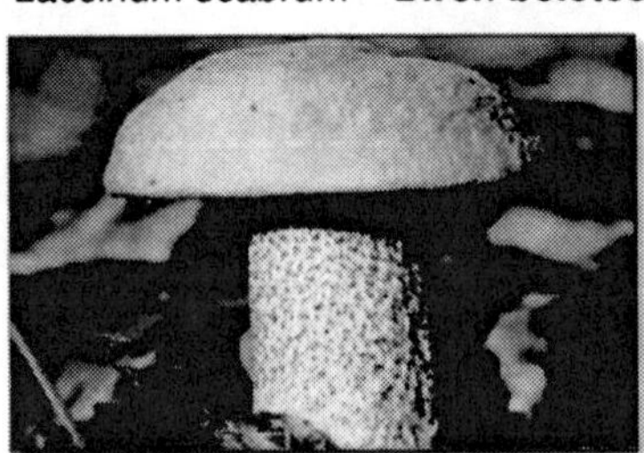

Cap colors vary from yellow to buff to orange to lilac, waxy, fibrous, stalk often twisted.

Season of fruiting: Summer fall

Ecosystem function: Mycorrhizal with aspen, spruce, and pine of all ages

Edibility: Good

Fungal note: Most common mushroom on upland site and was first mycorrhizal fungus to have its entire genome sequenced.

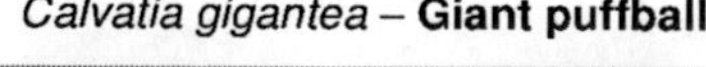

Calvatia gigantea – **Giant puffball**

Soft ball is soccer ball in size, white leathery skin when young turning yellow tan when mature

Season of fruiting: Late summer-fall

Ecosystem function: Mycorrhizal

Edibility: Edible when young stage

Fungal note: Giant puffballs 30.5 cm in dia can produce 7 trillion or more spore that are perfectly adapted to wind dissemination.in calm air spore fall at the rate 0.5mm per second

Polyporus alveolaris – **Diamond polypore**

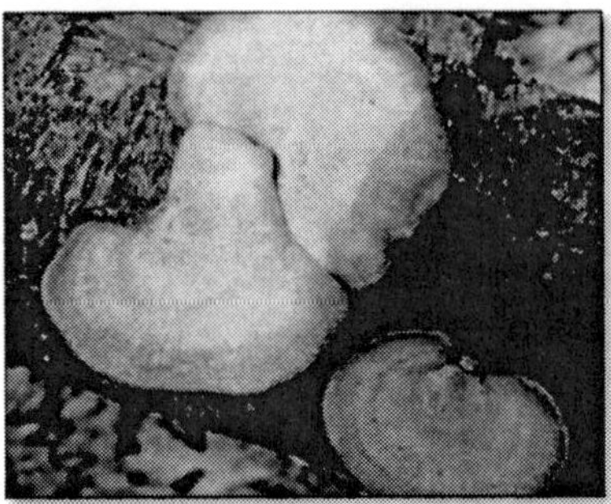

Fruit body cream to orange or reddish brown, short lateral stalk white to buff color, and large diamond shaped tubes

Season of fruiting: Spring early summer on dead hardwood branches

Ecosystem function: White rot

Edibility: Edible when young stage

Fungal note: Can cause decay when wood is at low moisture content

Boletus edulis – **The king boletes**

Cap cream brown to reddish brown, tube opening random resembling a sponge, flesh white yellow, stalk white ivory with fine lines reticulation forming a net

Season of fruiting: Late summer-fall

Ecosystem function: Mycorrhizal with pine, spruce, oak and birch

Edibility: Good

Fungal Note: Anti-cancer properties

Armillaria solidipes – **Honey Mushroom**

Cap golden yellow, prominent ring on stem, black shoe-string cords (rhizomorphs), under bark of infected trees or in the soil

Season of fruiting: Late summer-fall

Ecosystem function: Causes a root and butt rot of pine

Edibility: Edible

Fungal note: Species are luminescent often glowing in patches of decayed root or stem tissue.

Hydnum repandum – **Hedgehog mushroom**

Cap buff tan dull orange with white yellow teeth on underside

Season of fruiting: Summer-fall

Ecosystem function: Litter decay hardwood and conifer stand

Edibility: Edible

Fungal note: Spore produced on outside surface of the downward pointing teeth

Polyporus frondosus – **Hen of the wood**

Large dull white to gray, solitary fruit bodies with overlapping shelves on the ground near stumps or at the base of living hardwood trees

Season of fruiting: Late summer fall

Ecosystem function: White butt rot of hard woods

Edibility: Edible

Fungal note: Always found growing on the ground

7. Domestication and cultivation of wild edible mushroom in India

Case study 1: **An effort to domesticate wild edible mushrooms growing in the forest of Jharkhand** (Srivastava *et al.*, 2012)

Jharkhand has a rich biodiversity of wild edible mushrooms. A number of edible mushrooms growing in their natural habitats are being collected by the local people during the rainy season for their consumption or sale. Some of the common wild edible mushrooms of Jharkhand are *Macrolepiota procera*, *Termitomyces clypeatus*, *T. heimii*, *Lycoperdon*, *Calvatia*, *Geastrum*, *Boletus edulis*, *Russula*, *Termitomyces microcarpous*, *Amanita*, *Clitocybe*, *Armillaria* etc. They are rich in protein and can easily fit into all's platter, being a vegetarian product.

The villagers are acquainted with them, but they just collect them and consume. The idea to conserve and cultivate them is still eluding them. The ongoing study is an effort to domesticate some of these wild edible mushrooms in an artificial condition that they can be conserved and grown all throughout the year and standardize a package of practice for these mushrooms so that villagers could find some avenues to generate income through mushroom cultivation and marketing.

Selection of suitable substratum

The study of the natural habitats of the wild mushrooms to be domesticated reveals that these mushrooms grow in the soil enriched by the humus of the decaying leaves. Therefore to grow these mushrooms compost manure was prepared.In the compost leaves of *Shorea robusta, Syzygium cuminii, Mangifera indica, Bassia latifoloia*, and *Schleichera oleosa* were added as they are the main trees in the forest of Jharkhand. Due to lack facilities, we were not able preserved the spawn and not able to spawn on the compost prepared but it is our ongoing research. We will use readymade compost manure too which is used to cultivate already domesticated mushrooms. We have tried to identify them with their scientific names as well as their taxonomic position. Following are some of them

Case study 2: ***Tricholoma Giganteum*- A New Tropical Edible Mushroom for Commercial Cultivation in India** (Prakasam *et al.*, 2011)

In India, at present, four mushroom varieties viz., *Agaricus bisporus, Pleurotus* spp., *Volvariella* spp. and *Calocybe indica* have been recommended for the year round cultivation. The Indian subcontinent is known worldwide for its varied agro climatic ones with a variety of habitats that favour rich mushroom biodiversity. About two decades ago, *Calocybe indica* P. & C. was identified as a wild edible mushroom in India. . Only limited attempts were made for its cultivation until 1998. However, complete commercial production techniques were evolved for the first time in Tamil Nadu. *Tricholoma giganteum* Heim, a new edible mushroom pure white in colour

Collection of mushroom strains

During the year 2007, two milky (*Calocybe indica*) mushroom strains Ci (P) and Ci (N) and *Tricholoma* were collected from Erode and Coimbatore districts of Tamil Nadu. The mushrooms were pure cultured from the cap using tissue culture method and maintained on Potato Agar slants and used for further studies.

Growth and yield performance studies of wild mushroom strains

The Growth and yield performance studies of wild mushroom strains was conducted at the Mushroom Research and Training Centre, Department of Plant Pathology, Tamil Nadu Agricultural University, Coimbatore. A milky mushroom variety- *Calocybe indica* var. APK 2 was identified and released as a commercially cultivated species from Tamil Nadu Agricultural University, Coimbatore during 1998. It was used for comparison. The cultures of different strains of wild *C.indica* Ci (P) and Ci (N) collected from different places and *Tricholoma giganteum* were inoculated onto sorghum grains for spawn production.

Using the sorghum-based spawn, cylindrical beds were prepared using paddy straw as substrate (1 kg of paddy straw/bed) .Four to six holes were made on the sides of the beds for aeration. The beds were incubated at room temperature of 28+2°C for spawn running. After complete spawn run, the beds were cut into two equal halves and steamed casing soil (garden soil) was applied uniformly over the spawn run beds. The moisture level was maintained by regular water spraying on beds. The cased beds were incubated in partially sunken poly houses roofed with blue coloured high density polythene sheet, where a temperature of 30-35°C with 7580% relative humidity was maintained. The observations on Days for Spawn Run (DFSR), Days.

Performance testing of *T. giganteum* at farmer's location

This study was conducted in different farms at different locations under controlled conditions (with temperature range of 30-35°C with 75-80 % relative humidity) so as to test the yield performance and the consumers acceptability of the mushroom species. The results of yield trials conducted at Farm 1 indicated *T. giganteum* as the best performer based on maximum mean yield (820 g /bed) and bio efficiency (164%) compared to *C. indica* var. APK2 with 172% bioefficiency .Also, at Farm 2, *T.gignateum* performed with significantly higher yield of 847 g/ bed and bioefficiency of 169% compared to *C. indica* var. APK 2 (827 g/ bed; 165% bioefficiency) At Farm 3, T. *gignateum* recorded a yield (870 g/ bed; 174% bioefficiency) on par with *C. indica* var. APK 2 (Table 4). In all three trials, there was no variation in DFSR, DFPF and DFFH.

The nutritive values analyzed for *P. djamor* showed the presence of all essential nutrients with a calorific value of 19.8 Kcal/100 g of fresh mushrooms.

No incidence of pest and disease was recorded. The mushrooms could be stored under room temperature for two days and under refrigerated storage for 6 days without any microbial spoilage and liquefaction. The mushroom *T. giganteum* resembles milky mushroom in morphology. However, the stipe is sub-globose in *T. giganteum* where as in milky mushroom the stipe is elongated. The possibility of commercial cultivation of a new edible mushroom, *Tricholoma lobayense* closely resembling *C. indica* was reported

Outcome

Development of mushroom strains well adapted to the hot climatic plains of India with suitable Simpler cultivation technology, higher yield potential and prolonged shelf life are the present day needs of commercial cultivation. Being a tropical mushroom, *T. giganteum* has greater scope for commercial exploitation throughout the globe. The simple production techniques with sustainable yield, increased shelf life, attractive color, flavor and shape are the special features of this new edible mushroom.

Case study 3: **Mushroom in the food culture of the kaani tribe of kanyakumari district** (Sargunam *et al.*, 2012)

In the state of Tamil Nadu 36 tribal communities live and six of them are found in southern most kanyakumari district according to the census in1981. Kaani is a word denoting 24 manais while one mania is a piece of land. The staple food of the kaani tribe is rice; the tribal people were in the habitat of using a traditional wild strain of paddy called karainel. The traditional hunter gather instinct still exists in the tribal people and the collect mushroom in forest.

The tribal people collect the mushrooms early in the morning in bamboo or reed baskets. The collected mushrooms are cleaned in fresh water twice or thrice and slightly pounded in wooden mortar with an equal amount of rice. Then both are boiled with little water, spice, salts and wild green chillies are added for flavor and aroma. Then it`s served with cooked rice or cooked tapioca. Grated coconut is added to this preparation by some tribal people.

List of mushrooms collected by Kaani tribe

Botanical name	Vernacular name
Pleurotus sajor caju	Vellathazam Kumizh
Termitomyces heimii	Putru Kumizh
Termitomyces microcarpus	Ari Kumizh
Volvariella volvaceae	Uppu Kumizh
Auricularia auriculata	Murukkam Kumizh
Lentinus fusipes	Mozhaan Kumizh
Lentinus tuberegium	Kollamng Kumizh

8. Status and scenario of mushroom cultivation in India

Leading states of mushroom cultivation

Sl.No	States/Union Territory	Annual production(tons)
1	Punjab	60,000
2	Uttarakhand	8,000
3	Haryana	7178
4	Uttar Pradesh	7000
5	Tamil Nadu	6,500
6	Himachal Pradesh	5993
7	Orissa	5846
8	Andhra Pradesh	3022
9	Delhi	3010
10	Maharashtra	2975

(Manjith Singh *et al.*, 2010)

8.1. World production of fresh Mushrooms (2012)

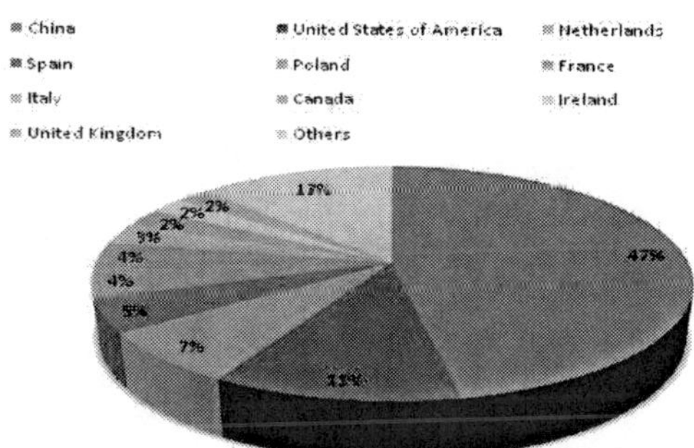

8.2. World Mushroom consumption by National Research Centre of Mushroom Report, 2012

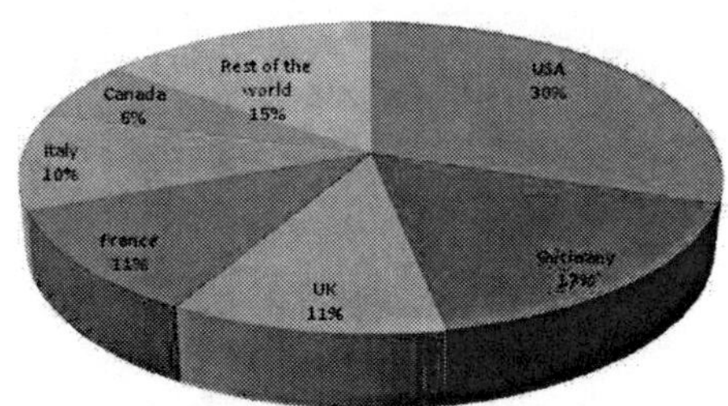

8.3. Nutritive value of Mushrooms (%)

Mushroom	Moisture	Protein	Fat	Fibre	CHO
Button mushroom	85.3	23.9	1,75	8.0	51.30
Oyster mushroom	90.1	26.6	2.00	13.3	50.7
Milky mushroom	85.6	32.3	0.70	41.0	59.8
Paddy straw mushroom	90.1	21.2	10.0	11.1	58.6
Shiitake mushroom	90.0	13.4	8.00	7.30	67.5
Jelly mushroom	89.1	8.3	8.20	4.70	63.0

8.4. Comparison of the nutritional index of different foods for mushrooms

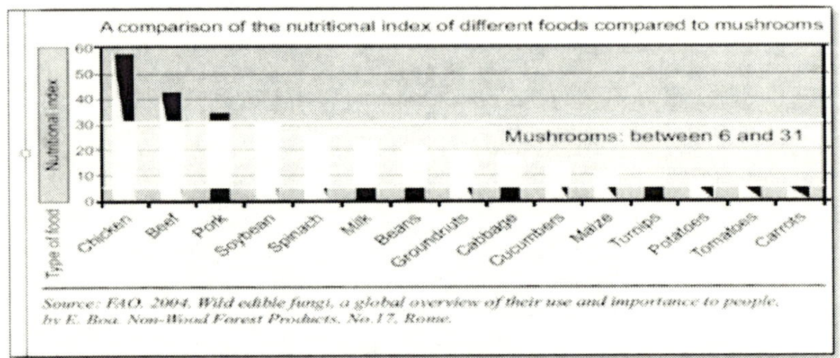

Source: FAO, 2004. Wild edible fungi, a global overview of their use and importance to people, by E. Boa. Non-Wood Forest Products, No.17, Rome.

Case study 4: **Nutritive Value of Edible Wild Mushrooms Collected from the Western Ghats of Kanyakumari District** (Johnsy *et al.*, 2012)

Mushroom	Moisture	Dry matter	CHO	Lipid	Protein	Fibre content
Pleurotus roseus	94.1	5.9	42.97	2.02	30.27	4.2
Pleurotus ostreatus	95.13	4.87	43.4	2.47	37.63	4.2
Pleurotus sajor caju	92.67	7.33	38.57	1.17	39.1	4.9
Termitomyces microcarpus	90.67	9.33	46.53	2.33	29.4	11.5
Termitomyces heimii	92.5	7.5	39.03	2.11	34.2	9.73
Auricularia auricula	95.17	4.83	33.23	1.63	36.3	8.4
Volvariella volvacea	90.67	9.67	43.53	2.04	30.57	9.67
Lentinus squarrosulus	90.5	11.03	47.83	2.58	37.13	11.33
Lentinus tuber regium	88.48	12.3	50.2	2.17	28.93	12.17
Grifola frondosa	87.3	9.4	40.77	1.49	31.47	7

The nutritional value of 10 edible mushroom species were analyzed, which forms a part of the food culture of the Kaani tribal community settled in the forests of Kanyakumari district in Tamil Nadu. Young and matured carpophores of 10 common wild edible mushrooms were collected from different locations in the Western Ghats of Kanyakumari district

People management

Although forests and fungi evolved for eons without human influence, people have played an integral role in shaping forest ecosystems since we began burning and clearing forests. Indeed, many edible mushroom species are found most abundantly in burned or young forests. Social, cultural, and economic values continue to determine forest use and management and are integral to addressing mushroom harvesting issues. The extensive cultural diversity of mushroom harvesters in the PNW-US is a salient example (Arora, 1999). Harvester groups include Native Americans, descendents of European settlers, and recent Latin American and Southeast Asian immigrants (Richards and Creasy, 1996; Liegel, 1998). Among and within each of these broad harvester categories are considerable differences in the way groups or individuals view their interactions with the forest and with each other.

These differences are frequently rooted in homeland cultural traditions and fundamental concepts, such as ownership, rights, freedoms, security, livelihoods, and spiritual connections with nature. Income is only one of many motivations for commercial harvesting; other reasons include independence (Sullivan, 1998), the excitement of the hunt, beautiful natural surroundings, and social activities in camps. Differences between recreational and commercial harvesters also have political and policy ramifications (McLain et al., 1998).On public lands, forest managers need to be sensitive to cultural nuances and differences if they are to simultaneously minimize conflicts and ensure resource conservation (Pilz et al., 1999). Some of the issues they face with all users include camping and sanitation facilities, traffic safety and road closures, fire danger, weapons, littering, wildlife disturbance, pathogen dispersal, impacts on other sensitive species or archeological sites, and conflicts among user groups, such as recreational harvesters, commercial harvesters, and big game hunters. Complete ecosystem management plans must incorporate economic information too.

Many forest associated rural communities have suffered economically and socially from reductions of timber harvesting on Federal lands. Development of NTFP enterprises can play a role in ameliorating these impacts, especially if companies can depend on reliable resource supplies from Federal lands to support local processing and value-added activities. Supplies are not the only salient issue, however. The volatility of international commerce in edible forest mushrooms (Blatner and Alexander, 1998) results in unpredictable demand for mushrooms. For example, the vast majority of matsutake are sold to Japan, but the mushroom commands such high prices that it is considered a luxury item.

Economic cycles or changes in monetary exchange rates can considerably alter prices paid to harvesters and their resulting incentive to collect. Harvest levels in other countries also vary from year to year, so global competition is unpredictable for all internationally-marketed species Managers are well-advised to develop programs and regulations that accommodate fluctuations in demand. Likewise, local NTFP harvesters and industries can stabilize their income by diversifying the products they harvest or market. As another example of economic issues, the value of annual mushroom harvesting has been used by conservation organizations as an argument for appealing timber sales.

Although clear cut harvesting will arrest the fruiting of ectomycorrhizal mushrooms for a decade or more and thinning can suppress fruiting, mushroom harvesting and timber growing can be compatible activities during much of a timber rotation. We lack adequate information for accurate assessments of discounted present net worth for mushrooms, but several illustrative scenarios comparing timber and mushroom values have been developed (Liegel,1998; Pilz et al., 1999) Using these methods, managers can modify assumptions or insert site-specific data tom analyze the economic consequences of local decisions. Importantly, the harvest of mushrooms benefits different individuals, companies, and customers than does timber harvesting, and many other NTFPs and amenities are derived from any given forest.

9. Drying and Preservation methods of wild edible mushrooms

Drying and preservation fresh wild edible mushrooms have a short period during which they can be eaten or consumed. Owing to their perishable nature, they quickly deteriorate, rot, or shrivel up. On questioning local people about this aspect, it was realized that they consumed large number of the mushrooms in fresh form and only a few are preserved after sun-drying, smoke drying or salting. Mushroom species such as *Geopora arenicola, Sepultaria sumneriana*, *Morchella* spp., *Pleurotus* spp., *Russula* sp. and *Sparassis* spp. are sun-dried in open and then stored in gunny bags, polythene bags or jars.

These hypogeous mushrooms are thoroughly washed in water to remove soil debris adhering to the apothecia, sun-dried, salted and then mixed with turmeric powder for enhancing shelf-life in storage and off-season consumption.

9.1. Wild Edible Mushrooms : Traditional Drying and Storage methods.

a) Drying of morels in open

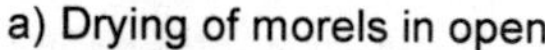

b) Fruiting bodies of *Geopora* spp. & *Sepultaria sumneriana* collected in baskets

c & d) Fruiting bodies of *Rhizopogon* spp. gathered from the forests

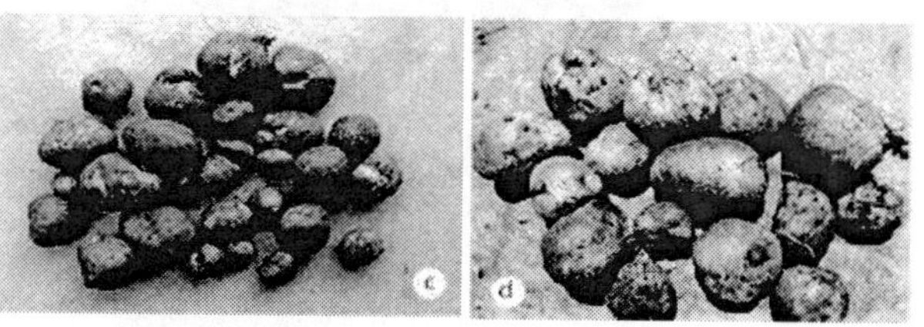

e) *Geopora arenicola* & *Sepultaria sumneriana* in 'Chajjh' for removing soil

f) Dried form of *Sparassis* spp. and morels

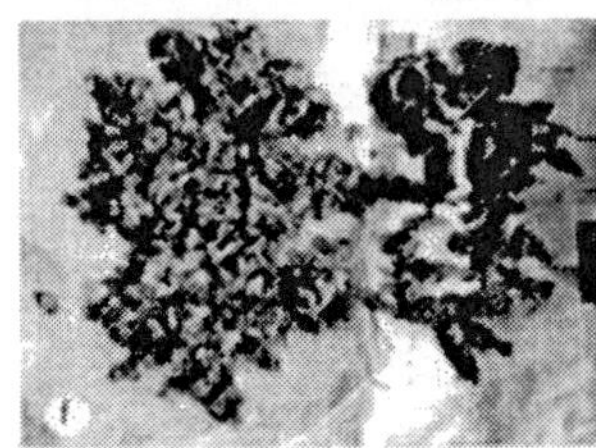

g) *Geopora arenicola* ready for cooking after proper washing

h) Morels packed in jars and polythene bags in a local market

10. Marketing and production of Wild edible mushroom in India

Case study 5: **Wild edible Macrofungal species consumed by the khasi tribe of Meghalaya, India** (Polashree Khaund and S R Joshi., 2013)

Wild edible mushroom collected from the forest by the ethnic tribes and sold in local markets of the khasi hills.The ethnic tribal population have extensive ethnomycological knowledge based on which they directly collect and sell the edible mushroom.11 different species were identified based on their morphology that belonged to 9 genera and 8 families. Clavulina spp was the most abundantly available species whereas albatrellus spp was already available in local markets.

Average cost of the wild edible mushroom in local market

Name	Price/Kg in INR
Gomphus floccosus	200
Tricholoma viridiolivaceum	200
Tricholoma sp.	200
Tricholoma saponaceum	200
Laccaria lateritia	280-300
Clavulina sp.	300-350
Ramaria	250-300
Craterllus odoratus	280
Cantharellus cibarius	280
Lactarius volemus	200
Albatrellus sp	200

Case study 6: **WEM in the livelihood of Santal Tribe in Sal forests of lateritic region of West Bengal** (Prakash Pradhanr *et al.*, 2009)

List of wild edible fungi used by Santals as food

The eastern part of India with its varied topography and climatic conditions harbors various forest types associated with different ethnic groups like the Santal, Lodha, Dhangar, etc. The undulating lateritic portion of this part of India is actually the

Chato Nagpur plateau extending from Bihar, Orissa, Chhattisgarh, and Jharkhand up to the western part of West Bengal The forests of the Lateritic region of West Bengal were found to be dominated by dry deciduous Sal (Family: Dipterocarpaceae). Other trees of the region included Arjuna (Terminalia arjuna (Roxb.

Different market price of wild edible mushroom (Rs/Kg) Rmp- Rampurhat, Ilm-Ilambazar, Snt-Sainthiya, Mln-Maladighi, Bsp-Bisnupur, Gji-Gangajalhati, Tal-Taldangra, Jhr-Jhargram.

WEM	Rmp	Ilm	Snt	Mln	Bsp	Gjl	Tal	Jhr
Amanita vaginata	50-70	40-50	50-80	40-50	55-80	40-50	40-55	40-55
Astraeus Hygrometricus	50-70	45-60	50-60	40-50	60-90	40-45	40-50	40-50
Russula delica	55-75	40-75	50-80	40-50	55-80	40-50	40-55	40-55
Termitomyces microcarpus	40-90	40-70	40-90	40-75	40-90	40-75	40-65	40-60
Pleurotus squarrosulus	60-100	60-80	-	60-70	-	60-70	60-80	60-75
Tricholoma crissum	-	60-90	70-150	-	75-130	60-80	60-90	70-85
Termitomyces heimii	-	70-200	-	50-120	70-160	50-100	40-100	50-110

Sustenance from wild edible mushrooms

The variety which had been exported in dried form i.e. Moral or Black mushrooms (*Morchella Spp*) commonly known as 'Guchhi' is collected as wild growth from coniferous forests of Himachal Pradesh, Jammu and Kashmir and Uttar Pradesh. Morels (gucchhi) growing only in wild, are the most valued wild mushroom in Western Europe particularly France, Germany, Italy and Switzerland. The international trade in dried morels is estimated to 225 tons annually. The suppliers are India, China, Turkey,

Pakistan, North America and Eastern Europe. Pakistan alone exports nearly 65 tons annually.

It is claimed that 2,89,000 persons are engaged annually in Pakistan in morel mushroom hunting on part time basis including 33% women, 27% men and 40% children from March to July months. The price for one kilogram dried morels is US $ 50 for the collector, $ 166 for wholesaler, $ 216 for exporter $ 330 for the importer. The global trade for yellow chanterelle mushroom (*Cantherellus* spp.) is much more lucrative than for morels, 200000 metric tons are bought and sold annually worldwide ranging from 1.25 to 1.4 billion US dollars every year. Germany is largest importer, followed by France and other Western Europe.

***Morchella* spp.**

In Himachal Pradesh and Jammu & Kashmir *Morchella* spp. are collected systematically during the growing seasons (spring and sometimes after rainy season) and sold to established markets both fresh and as dried mushrooms. In Madhya Pradesh, Chattisgarh and north eastern states wild mushrooms are sold in the local markets and provide sustenance to the tribal people and forest dwellers during the lean period (rainy season) when other non-wood forest products are not available in the forests. In the north-east of India some commonly wild mushrooms used in food are *Auricularia auricula, Schizophyllum commune, Lentinus* spp., etc. There is scope to tap the potential in and outside of our forests for providing sustained livelihood and profit to the people through systematic collection and processing of wild mushrooms, which is till today limited to the collection of morels only.

10.1. Food security

Mushrooms are of excellent food value as they provide a full protein food containing all the twenty one amino acids besides containing useful amount of fats, vitamins and minerals. Mushroom protein being easily digestible (70-90%) is considered superior to vegetable proteins. Two essential amino acids lysine and tryptophan are enormously present in mushrooms which are not found in cereals. Being low in caloric value (300 – 390 Kcal/100 g dry wit), low fat and high protein, they are considered as 'delight of diabetic patients'. Folic acid and Vitamin B-12 which are normally absent in vegetarian foods are present in mushrooms (3 g fresh mushroom can supply 1 micro g vitamin B12, recommended for daily uptake).

At present we have the lowest rate of protein consumption and due to population explosion the problem of protein hunger will become more acute. Under prevailing circumstances all possible sources of protein products will have to be exploited to save the country from hunger and malnutrition. Edible mushrooms can therefore be used as a weapon against starvation because of its high protein and vitamin content. And in a way contribute to food security by being easily available, affordable and usable.

11. Uses of wild edible mushroom

11.1. Uses Medicinal Values

Since thousands of years, edible fungi have been revered for their immense health benefits and extensively used in folk medicine. Specific biochemical compounds in mushrooms are responsible for improving human health in many ways. These bioactive compounds include polysaccharides, tri-terpenoids, low molecular weight proteins, glycoproteins and immunomodulating compounds.

Hence mushrooms have been shown to promote immune function; boost health; lower the risk of cancer; inhibit tumor growth; help balancing blood sugar; ward off viruses, bacteria, and fungi; reduce inflammation; and support the body's detoxification mechanisms. Increasing recognition of mushrooms in complementing conventional medicines is also well known for fighting many diseases.

1. **Good for heart**

The edible mushrooms have little fat with higher proportion of unsaturated fatty acids and absence of cholesterol and consequently it is the relevant choice for heart patients and treating cardiovascular diseases. Minimal sodium with rich potassium in mushroom enhances salt balance and maintaining blood circulation in human being. Hence, mushrooms are suitable for people suffering from high blood pressure. Regular consumption of mushrooms like *Lentinula, Pleurotus* spp decreases cholesterol levels.The lovastatin obtained from *Pleurotus ostreatus* and eritadenine obtained from shiitake has the ability to reduce blood cholesterol levels.

2. **Low calorie food**

The diabetic patients choose mushroom as an ideal food due to its low calorific value, no starch, little fat and sugars. The lean proteins present in mushrooms help to burn cholesterol in the body. Thus it is most preferable food for people striving to shed their extra weight.

3. **Prevents cancer**

Compounds restricting tumor activities are found in some mushrooms but only a limited number have undergone clinical trials. All forms of edible mushrooms, and white button mushrooms in particular, can prevent prostate and breast cancer. Fresh mushrooms are capable of arresting the action of 5-alpha-reductase and aromatase, chemicals responsible for growth of cancerous tumors. The drug known as Polysaccharide-K (Kresin), is isolated from *Trametes versicolor* (*Coriolus versicolor*), which is used as a leading cancer drug.

Such effects have been clinically validated in mushrooms like *Lentinula edodes, Trametes versicolor, Agaricus bisporus* and others. Selenium in the form of selenoproteins found in mushrooms has anticancer properties. According to the International Copper Association, the mushroom's high copper levels help to reduce colon cancer besides osteoporosis.

4. Anti-aging property

The polysaccharides from mushrooms are potent scavengers of super oxide free radicals. These antioxidants prevent the action of free radicals in the body, consequently reducing the aging process. Ergothioneine is a specific antioxidant found in *Flammulina velutipes* and *Agaricus bisporus* which is necessary for healthy eyes, kidney, bone marrow, liver and skin.

5. Regulates digestive system

The fermentable fiber as well as oligosaccharide from mushrooms acts as a prebiotics in intestine and therefore they anchor useful bacteria in the colon. This dietary fibre assists the digestion process and healthy functioning of bowel system.

6. Strengthens immunity

Mushrooms are capable of strengthening the immune system. A diverse collection of polysaccharides (beta-glucans) and minerals, isolated from mushroom is responsible for up-regulating the immune system. These compounds potentiate the host's innate (non-specific) and acquired (specific) immune responses and activate all kinds of immune cells. Mushrooms, akin to plants, have a great potential for the production of quality food. These are the source of bioactive metabolites and are a prolific resource for drugs. Knowledge advancement in biochemistry, biotechnology and molecular biology boosts application of mushrooms in medical sciences.

From a holistic consideration, the edible mushrooms and its by-products may offer highly palatable, nutritious and healthy food besides its pharmacological benefits. Still there are enough challenges ahead. Until now, how these products work is elusive and vast numbers of potential wild mushrooms are not explored. The utility of mycelia is paid little attention but it has tremendous potential, as it can be produced year around with defined standards. Knowledge on dose requirement, route and timing of administration, mechanism of action and site of activity is also lacking. Work is under progress in various laboratories across the world to validate these medicinal properties and to isolate new compounds.

If these challenges are met out in the coming days, mushroom industries will play a lead role in neutraceutical and pharmaceutical industries. The increasing awareness about high nutritional value accompanied by medicinal properties means that mushrooms are going to be important food item in coming days and at places may emerge as a substitute to non-vegetarian foods. Growing mushroom is economically and ecologically beneficial. Consuming mushroom is beneficial in every respect. Thus mushrooms are truly health food, a promising neutraceutical.

12. Constraints in wild edible mushrooms

As aforementioned, mushrooms with its huge health benefits can solve many a problems of under nutrition and malnutrition. Despite this fact mushroom cultivation and its utilization is not catching up fast.

The major constraints and how this can be prevailed over are:

1. Mushrooms are not popular in India. Both produces and consumers are not aware of its intrinsic worth. This can be overcome by organizing campaigns, training programmes, workshops, propagating its virtues through media etc.
2. Being highly perishable, lack of immediate access to markets is a major bottleneck in mushroom farming. The seasonality and wide fluctuation in collection results into erratic procurement and supply. The collection therefore may be organized by farming co-operatives or by NGOs/Traders.
3. As Mushrooms have to be immediately processed to increase its shelf life period, lack of storage facilities like processing units, cold storage, refrigerated transport etc are also one of the deterrents.

So it can be inferred that production of highly perishable commodities such as mushroom need a lot more than interim infrastructural facilities. It needs a synergy between various segments of cultivators, productions such as production centres, pre-cooling units, cold storages and export processing units or export processing zones and more so the ultimate consumers

13. Conclusion

The yields of wild edible fungi are the product of diverse and complex interactions within natural systems whose relations have coevolved over millennia. Managers cannot consider mushroom production in isolation from the community and ecosystem to which the mushrooms are ecologically adapted. Our approach to the harvest of wild edible mushrooms considers plants, the soil community, the larger ecosystem and human uses and needs as a coherent and dynamic partnership. The sustainability of the wild edible mushroom resource ultimately is understood in terms of patterns arising from the connections within the partnership.

Just as the forest invests tremendous capital in the form of photosynthates to fuel the production of wild edible mushrooms, we must invest in the effort to understand and conserve this ephemeral and poorly understood resource. Monitoring is essential and three kinds are recommended: detection, evaluation, and research monitoring of the mushroom resource are necessary to assess abundance and distribution, the effects of management, and the role of these fungi in long term forest health. Our biological research must be done in conjunction with socioeconomic questions relating to effects on rural communities, interactions between mushroom-user groups, and sustainability of a commercial industry. Sharpening our knowledge will be a good investment and is an essential part of maintaining continuous production of wild edible mushrooms for future generation.

Reference

Adhikari, M. K., Devkota, S. and Tiwari, R. D. 2005. Ethnomycolgical Knowledge on Uses of Wild Mushrooms in Western and Central Nepal. **Our Nature**. 3: 13-19.

Ajay K Srivastava, Fr Prabhat., Kennedy Soreng s.j 2012 An effort to domesticate wild edible mushrooms growing in the forest of Jharkhand

Allce, A. and M. Muthusamy. 1999. **Mushroom culture** p. 47.

Bahl, Nita, Medicinal Value of Edible fungi In: Proceeding of Te International conference on mushroom diversity. Macrofungi associated with oaks of eastern North America. Morgantown, WV: West VirginiaUniversity Press. 467 p.

Chandra, Aindrila, 1985. Manual of Indian Edible Mushrooms, Today & Tomorrow's Printers and Publishers, New Delhi, 267 p.

Davidson Sargunam.S.G., Johnsy, A.Selvin Samuel and v.kaviyarasan.2012.**Indian Journal of Traditional knowledge**, 11(1):150-153

Dyke, A. J. and Newton, A. C. 1999. Commercial harvesting of wild mushrooms in Scottish forest: is it sustainable? **Scottish forestry**. 53:77-85.

Eric boa.2004. Wild Edible Fungi. A Global Overview of Their Use and Importance to People.**FAO**.Rome

Fogel, Robert. 1980. Mycorrhizae and nutrient cycling in natural forest ecosystems. Forests: implications for preserving biodiversity in managed stands. **BioScience**. 41: 382-392.

Hansen, A.; Spies, T.A.; Swanson, F.J.; Ohmann, J.L. 1995. Lessons from natural mushroom. **BioScience**. 47: 282-287.

Maha, M. and D. S. Pangerteni. 1989. Pengawetan jamur merang (*Volvariella volvaceae*) dengan kombinasi pemanasan dan iradiasi *In* Risalah Simposium IV Aplikasi Isotop dan Radiasi, Jakarta, 13-15 Des. 1207-1215.

Manjit singh., Bhuvnesh Vijay and Shwet Kamal.2011.Mushroom Marketing, Cultivation and Consumption.ICAR.Solan. **New Phytologist**. 86:199-212.

Michel E.ostry., Neil A.Anderson and Joseph G.O`Brien.2010.Field Guide to Common Macrofungi in Eastern Forest and their Ecosystem Function.**USDA**

Muraleedharan TR, Iyengar L and Venkobachar C.1995. Screening of Tropical Wood-Rotting Mushroom for Copper Biosorption, **Appl Env microbiol** 61 (9): 3507.

Pilz D and Molina R. 2001. Commercial harvests of Edible Mushrooms from the forests of the Pacific Northewst United States: Issues, Management and Monitoring for sustainability. **Forest Ecology and management**: 1-14.

Polashree khaund and S.R.Joshi.2013.**Indian Journal of Natural Products and Resources**, 4(2):197-204

Prakash Pradhanr, Subhankar Banerjeer, Anirban Roy and Krishnendu Acharya Oct 2009) WEM in the livelihood of Santal Tribe in Sal forests of lateritic region of West Bengal.

Purkayastha, R. P. and Aindrila Chandra, 1976. Indian Edible Mushrooms, Firma K.L.M. Calcutta science and cultivation technology of edible fungi (Kaul and Kapur Ed) pp. 204-209 (1983).

Sachan,S.K.S.,Patra,J.K.and Thatoi,H.N.2013.Journal of Agricultural Technology Indigenius Knowledge of Ethinic Tribes for Utilization of Wild Mushroom As Food and Medicine in Similipal Biosphere Reserve,Odisha,India ,Vol 9(2):403-416

Sharma, B.M. Singh B.M. and Kumar, A., Notes on some promising wild Edible Mushroom Shu-Ting Chang and Philip G. Miles, 2004. Mushrooms - Cultivation, Nutritional Value, Medicinal Effect and Environmental Impact, Second Edition, CRC Press, 451 p.

Upadhyay, R. C., Kaur, A., Kumari, D., Scmwal, K. C. and Gulati, A. 2008. Collection of rich varieties of wild edible mushroom from north Himalaya India. **International J. Agriculture and Biology**. 38:24-39.

CPSIA information can be obtained at www.ICGtesting.com
Printed in the USA
LVOW10s1332180516

488844LV00002B/154/P

9 783659 518034